Name _____

1. $\frac{4}{7} - \frac{2}{7} =$

   ○ $\frac{6}{7}$ ○ $\frac{2}{7}$

   ○ $\frac{2}{0}$ ○ $\frac{6}{14}$

2.   31.6
     +14.8

   ○ 49.1   ○ 45.4

   ○ 26.8   ○ 46.4

3. 2)$\overline{192}$

   ○ 96   ○ 86

   ○ 906   ○ 90 remainder 1

4. 3 quarters + 4 dimes + 3 nickels =

   ○ $1.05   ○ $1.20

   ○ $1.30   ○ $1.18

Name _____   5E-2

1. Reduce (simplify) to lowest terms:

$$\frac{2}{4}$$

○ $\frac{4}{8}$ ○ $\frac{1}{2}$

○ 1   ○ 6

2. 200
   x 14

○ 2954   ○ 1000

○ 3800   ○ 2800

3. 668
   -307

○ 301   ○ 975

○ 261   ○ 361

4. Find the missing number.

24, 20, 16, ___, 8

○ 10   ○ 12

○ 14   ○ 18

Name _____   5E-3

1. 248
    768
    +338
    ───

    ○ 1356   ○ 1344

    ○ 1354   ○ 1334

2. 417
    x 2
    ────

    ○ 824   ○ 419

    ○ 834   ○ 934

3. 6000 + 400 + 30 + 1 =

    ○ 6,431   ○ 60,431

    ○ 64,031   ○ 6,341

4. Reduce (simplify) to lowest terms:

    $$\frac{3}{6}$$

    ○ $\frac{1}{2}$ ○ 9

    ○ $\frac{6}{12}$ ○ $\frac{1}{3}$

Name _____   5E-4

1. 25.8
   +13.8

   ○ 38.6   ○ 39.6

   ○ 49.6   ○ 12.0

2. Find the average of these numbers:

   (20, 30, 40)

   ○ 90   ○ 20

   ○ 30   ○ 40

3. An office building has 20 floors. The elevator stopped on the fifteenth floor. How many floors are above the elevator?

   ○ 20   ○ 5

   ○ 10   ○ 4

4. 5)215

   ○ 53   ○ 403

   ○ 43   ○ 40 remainder 5

Name _____   5E-5

1. ___ x 5 = 35

　○ 40　　　　○ 7

　○ 175　　　○ 9

2. Reduce (simplify) to lowest terms:

$$\frac{4}{6}$$

　○ $\frac{2}{3}$　　　　○ 2

　○ $\frac{1}{2}$　　　　○ 10

3. 6 x 4 = 4 x ___

　○ 10　　　　○ 14

　○ 24　　　　○ 6

4.   87,650
　+49,637

　○ 137,287　　○ 137,280

　○ 136,287　　○ 18,397

Name _____ 5E-6

1. 262
    x 34

    ○ 1834   ○ 8908

    ○ 7908   ○ 7708

2. $81.38
    -70.17

    ○ $11.21   ○ $151.55

    ○ $10.21   ○ $151.21

3. 3 thousands
    4 hundreds
    7 tens
    5 ones =

    ○ 19   ○ 5,743

    ○ 3,475   ○ 30,475

4. 36.3
    -14.8

    ○ 51.1   ○ 22.5

    ○ 21.5   ○ 50.5

Name _____ 5E-7

1. 4)̄408

    ○ 24   ○ 102

    ○ 12   ○ 92

2. 6290
    -4846

    ○ 2,656   ○ 1,444

    ○ 2,454   ○ 11,136

3. Reduce (simplify) to lowest terms:

    $\frac{2}{8}$

    ○ 6   ○ $\frac{1}{4}$

    ○ $\frac{1}{2}$ ○ 10

4. Which fraction tells how much of the figure below is shaded?

    ○ $\frac{2}{4}$ ○ $\frac{2}{3}$

    ○ $\frac{1}{3}$ ○ $\frac{1}{2}$

Name _____

1.
$$\frac{1}{4}$$
$$+\frac{1}{4}$$

○ $\frac{2}{8}$ ○ $\frac{1}{2}$

○ $\frac{0}{4}$ ○ $\frac{1}{16}$

2. George Washington became P[resident] in 1789. He was 57 years o[ld] what year was he born?

○ 1732   ○ 1846

○ 1722   ○ 1022

3. Round 79 to the nearest ten.

○ 70   ○ 90

○ 80   ○ 78

4.
794
+ 56

○ 740   ○ 851

○ 850   ○ 738

Name _____   5E-9

1. $6.55
   -1.38

   ○ $5.23   ○ $5.27

   ○ $7.93   ○ $5.17

2. $\frac{3}{8}$
   $+ \frac{1}{8}$

   ○ $\frac{1}{4}$ ○ $\frac{1}{2}$

   ○ $\frac{3}{64}$ ○ $\frac{1}{3}$

3. 351
   x 45

   ○ 15,795   ○ 3,159

   ○ 396   ○ 13,795

4. $\frac{1}{3}$ $\frac{1}{4}$ ⬭$\frac{1}{8}$⬭

   Of the three fractions shown, the one that is circled has:

   ○ the greatest value
   ○ the least value
   ○ the same value

Name _____   5E-10

1. Abraham Lincoln became President in 1861. He was 52 years old. In what year was he born?

　○ 1809　　○ 1913

　○ 1811　　○ 1709

2.　　436
　　−347

　○ 111　　○ 99

　○ 783　　○ 89

3.　4)605

　○ 158　　○ 151 remainder 1

　○ 155　　○ 121 remainder 1

4.　$\frac{2}{6}$
　$+\frac{2}{6}$

　○ $\frac{2}{3}$　　○ $\frac{1}{3}$

　○ $\frac{1}{9}$　　○ $\frac{0}{0}$

Name _____   5E-11

1. $\frac{4}{6}$
    $-\frac{1}{6}$

    ○ $\frac{1}{2}$ ○ $\frac{1}{3}$

    ○ $\frac{5}{6}$ ○ $\frac{1}{6}$

2. 340
    x 2

    ○ 680   ○ 682

    ○ 780   ○ 342

3. 3 x (2 x 5) = (3 x ___) x 5

    ○ 10   ○ 2

    ○ 7   ○ 5

4. 58.1
    -14.6

    ○ 44.5   ○ 72.7

    ○ 33.5   ○ 43.5

Name _____   5E-12

1. $58.30
   + 9.48
   ─────

   ○ $67.40   ○ $67.78

   ○ $57.78   ○ $48.82

2. $\frac{4}{8}$
   $-\frac{2}{8}$
   ─────

   ○ $\frac{1}{4}$ ○ $\frac{1}{2}$

   ○ $\frac{3}{4}$ ○ $\frac{1}{8}$

3. 154
   x 16
   ─────

   ○ 2374   ○ 2464

   ○ 2364   ○ 1078

4. Which sign goes in the circle below?

   30 ÷ 5 ◯ 6 x 9

   ○ >
   ○ <
   ○ =

Name _____   5E-13

1. $3.30
      x   2

     ◯ $3.32   ◯ $7.60

     ◯ $6.62   ◯ $6.60

2. $\frac{3}{7} - \frac{2}{7} =$

     ◯ $\frac{5}{7}$ ◯ $\frac{1}{0}$

     ◯ $\frac{1}{7}$ ◯ $\frac{5}{14}$

3. $3\overline{)904}$

     ◯ 301 remainder 1
     ◯ 311 remainder 1
     ◯ 301
     ◯ 234 remainder 2

4.   506
     +396

     ◯ 892   ◯ 902

     ◯ 802   ◯ 110

Name _____ 5E-14

1. $\frac{4}{8}$
    $-\frac{2}{8}$

    ○ $\frac{1}{2}$ ○ $\frac{1}{4}$

    ○ $\frac{2}{3}$ ○ $\frac{3}{8}$

2. 3000
    -2679

    ○ 321   ○ 1679

    ○ 1321   ○ 5679

3. Mrs. Baker started making a quilt on April 20th. She finished it 4 months later. In what month did she finish the quilt?

    ○ August   ○ July
    ○ September   ○ June

4. <u>0</u> hundreds <u>6</u> tens <u>4</u> ones = _____

    ○ 604   ○ 64

    ○ 46   ○ 406

Name _____   5E-15

1. 60.7
   +28.7

   ○ 48.0   ○ 88.4

   ○ 88.14   ○ 89.4

2. Round 851 to the nearest hundred.

   ○ 800   ○ 900

   ○ 100   ○ 850

3. 8 dimes + 5 nickels + 2 pennies =

   ○ $1.07   ○ $1.17

   ○ $.97   ○ $1.27

4. 236
    x 32

   ○ 1180   ○ 7552

   ○ 7652   ○ 7542

Name _____ 5E-16

1. It is 3:15 in California. In New York it is 3 hours later. What time is it in New York?

   ○ 1:15   ○ 6:45

   ○ 12:15   ○ 6:15

2. $\dfrac{3}{4}$
   $-\dfrac{1}{4}$

   ○ $\dfrac{1}{2}$ ○ 1

   ○ $\dfrac{1}{4}$ ○ $\dfrac{3}{8}$

3. $6\overline{)189}$

   ○ 31 remainder 3
   ○ 63
   ○ 313 remainder 3
   ○ 301 remainder 3

4. $86.48
   +73.21

   ○ $160.70   ○ $13.27
   ○ $159.69   ○ $160.69

Name _____ 5E-17

1. 18 ÷ 2 =

    ○ 20   ○ 9

    ○ 36   ○ 8

2. 47.4
    -31.9

    ○ 15.5   ○ 16.5

    ○ 79.3   ○ 16.3

3. 60,009
    +57,306

    ○ 110,003   ○ 2,703

    ○ 117,305   ○ 117,315

4. An airplane can fly 400 miles an hour. How far can it fly in 3 hours?

    ○ 403 miles   ○ 1200 miles

    ○ 1233 miles   ○ $133\frac{1}{3}$ miles

Name _____ 5E-18

1. $\frac{5}{10}$
    $+\frac{3}{10}$

    ○ $\frac{4}{5}$ ○ $\frac{1}{5}$

    ○ $\frac{1}{10}$ ○ $\frac{3}{5}$

2. 8273
    -1619

    ○ 6654   ○ 6664

    ○ 7466   ○ 9892

3. 5 ft. 10 in.
    +3 ft. 6 in.

    ○ 9 ft. 6 in.
    ○ 9 ft. 4 in.
    ○ 2 ft. 4 in.
    ○ 9 ft. 16 in.

4. What number has the greatest value?

    ○ 5102   ○ 5201

    ○ 5012   ○ 2015

Name _____ 5E-19

1. 38.5
   +40.5

   ○ 78.5   ○ 79.0

   ○ 89.0   ○ 78.0

2. In New York it is 1:00. In California it is 3 hours earlier. What time is it in California?

   ○ 10:00   ○ 4:00

   ○ 11:00   ○ 9:00

3. 9 ÷ 3 =

   ○ 27   ○ 12

   ○ 6   ○ 3

4. 240
   x 40

   ○ 960   ○ 9600

   ○ 9640   ○ 280

Name _____   5E-20

1.   79
     x11

   ○ 90   ○ 158

   ○ 879   ○ 869

2.   6)̄608

   ○ 101 remainder 2
   ○ 111 remainder 2
   ○ 108
   ○ 991 remainder 8

3.   There were 12 ice cream bars in the freezer. Half of them have been eaten. How many are left?

   ○ 6   ○ 4
   ○ 8   ○ 10

4. $\frac{6}{10}$
     $-\frac{4}{10}$

   ○ $\frac{1}{10}$ ○ $\frac{1}{5}$

   ○ 1   ○ $\frac{2}{5}$

Name _____   5F-1

1. 72.8
   +15.9

   ○ 87.7   ○ 88.7

   ○ 56.9   ○ 121.1

2. Reduce (simplify) to lowest terms:

   $$\frac{2}{8}$$

   ○ $\frac{1}{4}$ ○ $\frac{1}{2}$

   ○ 4   ○ $\frac{2}{4}$

3. 4)‾848‾

   ○ 202   ○ 21 remainder 8

   ○ 212   ○ 2012

4. $5.00
   x   5

   ○ $52.00   ○ $25.00

   ○ $5.50   ○ $30.00

Name _____ 5F-2

1. $\frac{5}{6}$
    $-\frac{3}{6}$

    ○ $\frac{1}{3}$ ○ $1\frac{1}{3}$

    ○ $\frac{1}{6}$ ○ $\frac{1}{2}$

2. 500
    × 25

    ○ 3,500   ○ 12,000

    ○ 12,500   ○ 12,775

3. 591
    −401

    ○ 101   ○ 190

    ○ 992   ○ 191

4. $9.90
    − 4.68

    ○ $5.38   ○ $14.58

    ○ $5.32   ○ $5.22

Name _____   5F-3

1. 165
    x 4

    ○ 760   ○ 822

    ○ 660   ○ 680

2. Find the average of these numbers:
    (6, 2, 4, 8, 5)

    ○ $8\frac{1}{3}$ ○ 4

    ○ 6   ○ 5

3. $33.46
    21.58
    +26.92

    ○ $81.96   ○ $111.21

    ○ $81.18   ○ $82.06

4. 56.8
    -12.9

    ○ 43.9   ○ 33.9

    ○ 44.1   ○ 69.7

Name _____ 5F-4

1. $1\frac{1}{5}$
    $+2\frac{1}{5}$

    ○ $3\frac{1}{2}$ ○ $3\frac{2}{5}$

    ○ 3 ○ $3\frac{1}{10}$

2. The factors (divisors) of 4 are:

    ○ 1, 2, 4   ○ 0, 1, 2, 4

    ○ 0, 1, 4   ○ 4, 8, 12

3. If a car goes 16 miles on one gallon of gas, how far can it go on 8 gallons of gas?

    ○ 128 miles   ○ 24 miles

    ○ 2 miles   ○ 104 miles

4. $4\overline{)416}$

    ○ 114   ○ 1004

    ○ 104   ○ 914

Name _____   5F-5

1. 326
    x 23

    ○ 7498   ○ 1630

    ○ 9721   ○ 7588

2.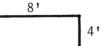

    What is the perimeter (distance around) the rectangle above?

    ○ 32'   ○ 24'

    ○ 12'   ○ 28'

3. The factors (divisors) of 6 are:

    ○ 6, 12, 18   ○ 0, 1, 2, 3

    ○ 1, 2, 3, 6   ○ 1, 6

4. $3 \frac{1}{8}$
    $+4 \frac{2}{8}$

    ○ $7 \frac{1}{2}$ ○ $7 \frac{1}{4}$

    ○ $7 \frac{3}{16}$ ○ $7 \frac{3}{8}$

Name _____ 5F-6

1. Find the number with an 8 in thousands place.

   ○ 81,702   ○ 78,201

   ○ 72,801   ○ 73,486

2. 34.63
   +31.48
   ──────

   ○ 66.11   ○ 66.10

   ○ 76.11   ○ 3.15

3. $4\frac{1}{7}$
   $+3\frac{1}{7}$
   ─────

   ○ $7\frac{1}{7}$ ○ $1\frac{0}{7}$

   ○ $7\frac{2}{7}$ ○ $7\frac{1}{3}$

4. The factors (divisors) of 8 are:

   ○ 2, 4, 8   ○ 1, 2, 4, 8

   ○ 0, 2, 4, 8   ○ 8, 16, 24

Name _____ 5F-7

1. 10,760
    - 9,485

    ○ 1,325   ○ 1,270

    ○ 20,245   ○ 1,275

2. $8 \frac{1}{8}$
    $+2 \frac{2}{8}$

    ○ $10 \frac{3}{8}$ ○ $10 \frac{1}{4}$

    ○ $6 \frac{3}{8}$ ○ $\frac{3}{8}$

3. $4\overline{)805}$

    ○ 201
    ○ 201 remainder 1
    ○ 211 remainder 1
    ○ 20 remainder 4

4. The factors (divisors) of 5 are:

    ○ 5, 10, 15   ○ 1, 5

    ○ 0, 1, 5   ○ 1, 2, 5

Name _____ 5F-8

1. Father bought a dozen eggs. He cooked 8 of them for breakfast. How many are left?

   ○ 4   ○ 8

   ○ 2   ○ 0

2.   56.8
     -12.9

   ○ 44.1   ○ 44.9

   ○ 69.7   ○ 43.9

3. Reduce (simplify) to lowest terms:

   $$\frac{3}{9}$$

   ○ $\frac{2}{3}$ ○ $\frac{1}{3}$

   ○ 3   ○ $\frac{1}{2}$

4. 4000 + 20 + 5 =

   ○ 4250   ○ 4205

   ○ 4025   ○ 425

Name _____   5F-9

1. 567
    x 45

    ○ 25,515   ○ 5,103

    ○ 612   ○ 25,615

2. 8 x 3 = ___ x 8

    ○ 11   ○ 3

    ○ 24   ○ 32

3. Which fraction tells how much
    of the figure below is shaded?

    ○ $\frac{1}{4}$ ○ $\frac{2}{3}$

    ○ $\frac{3}{4}$ ○ $\frac{3}{5}$

4. $1 \frac{1}{4}$
    $+2 \frac{1}{4}$

    ○ $3 \frac{3}{4}$ ○ $4 \frac{1}{4}$

    ○ $3 \frac{1}{2}$ ○ $\frac{1}{2}$

Name _____   5F-10

This graph shows the average monthly rainfall for Miami, Florida. Use the graph to answer the questions.

1. In which month is the average rainfall highest in Miami?
   ○ May   ○ July
   ○ September   ○ October

2. About how many inches of rain are there in October?
   ○ 9   ○ 8
   ○ 7   ○ 6

3. In what months is the rainfall about 7 inches?
   ○ Apr., May   ○ May, July
   ○ Aug., Oct.   ○ May, June

4. About how many inches greater is the rainfall in September than in February?
   ○ 7"   ○ 6"
   ○ 5"   ○ 4"

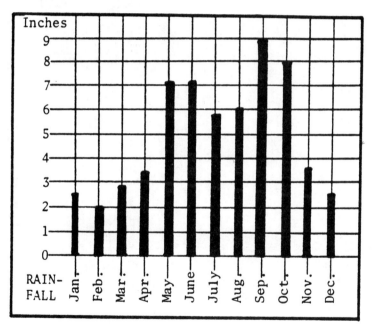

Name _____ 5F-11

1. $3 \frac{4}{6}$
    $+1 \frac{1}{6}$

    ○ 5   ○ $4 \frac{2}{3}$

    ○ $4 \frac{5}{6}$ ○ $2 \frac{1}{2}$

2. Which symbol goes in the circle below?

    $\frac{1}{2}$ ◯ $\frac{1}{3}$

    ○ >
    ○ <
    ○ =

3. $5\overline{)510}$

    ○ 102   ○ 99 remainder 4
    ○ 100   ○ 12

4. 56.79
    +12.56

    ○ 44.23   ○ 69.25
    ○ 68.216   ○ 69.35

Name _____   5F-12

1. $49.03
    -21.67

    ○ $27.36   ○ $27.37

    ○ $28.64   ○ $27.46

2. 1 = □/3

    ○ 1   ○ 3

    ○ 2   ○ 4

3.   605
    x 23

    ○ 3,025   ○ 13,915

    ○ 14,145   ○ 628

4. Round 51 to the nearest ten.

    ○ 50   ○ 60

    ○ 70   ○ 40

Name _____   5F-13

1. $1 = \dfrac{\square}{4}$ 2.   5000
                                                    −4891

   ○ 1   ○ 3   ○ 1891   ○ 9891

   ○ 2   ○ 4   ○ 109   ○ 1009

3. The factors (divisors)   4. $2\frac{1}{8}$
    of 10 are:
                                        $+3\frac{1}{8}$

   ○ 0, 1, 2, 5   ○ 10, 20, 30

   ○ 2, 5, 10   ○ 1, 2, 5, 10   ○ $5\frac{1}{4}$ ○ $5\frac{1}{16}$

                                     ○ $5\frac{1}{8}$ ○ $\frac{1}{4}$

Name _____   5F-14

1. 6)618̄

    ○ 13   ○ 103
    ○ 113   ○ 102

2. □/5 = 1

    ○ 2   ○ 3
    ○ 4   ○ 5

3. 48.0
    -16.9
    ─────

    ○ 32.1   ○ 32.9
    ○ 31.1   ○ 64.9

4. $68.93
    +61.46
    ─────

    ○ $7.53   ○ $132.19
    ○ $130.39 ○ $129.39

Name _____   5F-15

1. 244
    × 3

    ○ 732   ○ 247

    ○ 622   ○ 1011

2. Reduce (simplify) to lowest terms:

    $\frac{4}{8}$

    ○ $\frac{1}{2}$ ○ 2

    ○ $\frac{1}{4}$ ○ $\frac{2}{8}$

3. $4\frac{2}{6}$
    $+3\frac{2}{6}$

    ○ $7\frac{3}{6}$ ○ $7\frac{1}{3}$

    ○ $7\frac{1}{2}$ ○ $7\frac{2}{3}$

4. What is the missing number?

    24, 30, 36, ____, 48

    ○ 42   ○ 44

    ○ 40   ○ 46

Name _____   5F-16

1. 4869
    −1036

    ○ 3033   ○ 3833
    ○ 5905   ○ 3823

2. $\dfrac{\square}{2} = 1$

    ○ 1   ○ 3
    ○ 2   ○ 4

3. $\dfrac{2}{9} + \dfrac{1}{9} =$

    ○ $\dfrac{1}{6}$ ○ $\dfrac{1}{3}$

    ○ $\dfrac{2}{3}$ ○ $\dfrac{3}{18}$

4. 4 × (7 × 6) = (4 × ___) × 6

    ○ 13   ○ 28
    ○ 7   ○ 10

1. $3\frac{2}{3}$
   $-1\frac{1}{3}$

   ○ 5   ○ $1\frac{2}{3}$
   ○ $2\frac{1}{3}$ ○ 2

2. $\frac{3}{4}$
   $+\frac{1}{4}$

   ○ $\frac{1}{2}$ ○ 1
   ○ $1\frac{1}{4}$ ○ $\frac{5}{4}$

3. 234.16
   +185.48

   ○ 319.54   ○ 151.32
   ○ 419.91   ○ 419.64

4. Song bought 3 pounds of oranges for 20¢ a pound. How much change should he receive from a dollar?

   ○ 60¢   ○ 20¢
   ○ 80¢   ○ 40¢

Name _____   5F-18

1. $\frac{2}{8}$
    $+\frac{6}{8}$

    ○ 1   ○ $\frac{3}{4}$

    ○ $\frac{1}{8}$ ○ $1\frac{1}{8}$

2. 360
    × 30

    ○ 1,080   ○ 10,830

    ○ 10,800   ○ 390

3. The factors (divisors) of 3 are:

    ○ 1, 3, 6   ○ 0, 1, 3

    ○ 1, 3   ○ 3, 6, 9

4. $4\frac{3}{4}$
    $-1\frac{2}{4}$

    ○ $2\frac{1}{4}$ ○ $6\frac{1}{4}$

    ○ $3\frac{1}{8}$ ○ $3\frac{1}{4}$

Name _____   5F-19

1. 302
    x 3

    ○ 305   ○ 906

    ○ 916   ○ 936

2. $\frac{4}{5}$
    $+ \frac{1}{5}$

    ○ $1\frac{1}{5}$ ○ $\frac{6}{5}$

    ○ 1   ○ $\frac{1}{2}$

3. 45.26
    -13.20

    ○ 32.00   ○ 31.06

    ○ 58.46   ○ 32.06

4. 5)1007

    ○ 2001 remainder 2
    ○ 21 remainder 2
    ○ 201 remainder 2
    ○ 210 remainder 2

5F-20

1. $5 \frac{3}{4}$
    $-1 \frac{1}{4}$
    ───

    ○ $4 \frac{1}{2}$ ○ 7

    ○ $4 \frac{1}{4}$ ○ $3 \frac{1}{2}$

2. Anna had a board 12 feet long. She cut it into 3 equal pieces. How long was each piece?

    ○ 3 feet   ○ 4 feet

    ○ 36 feet   ○ 4 inches

3. 33,469
    -21,579
    ──────

    ○ 11,880   ○ 12,110

    ○ 55,048   ○ 11,890

4. $\frac{2}{3}$
    $+ \frac{1}{3}$
    ───

    ○ 1   ○ $\frac{4}{3}$

    ○ $\frac{1}{3}$ ○ $\frac{2}{3}$

Name _____   5G-1

1. $4.21
    x   3

   ○ $12.93   ○ $12.63

   ○ $4.24   ○ $12.73

2. $5 \frac{1}{4}$
   $+2 \frac{1}{4}$

   ○ $7 \frac{1}{8}$ ○ $\frac{1}{2}$

   ○ $7 \frac{1}{2}$ ○ 3

3. 8.6 − 3.5 =

   ○ 5.1   ○ 12.1

   ○ 4.9   ○ 4.1

4. $6\overline{)612}$

   ○ 12   ○ 102

   ○ 112   ○ 98

Name _____ 5G-2

1. 324
   -116

   ○ 212   ○ 108
   ○ 208   ○ 440

2. Three multiples of 3 are:

   ○ 1, 3, 6   ○ 3, 6, 9
   ○ 1, 2, 3   ○ 0, 1, 3

3. $\frac{1}{4} + \frac{1}{4} =$

   ○ $\frac{2}{8}$ ○ 1
   ○ $\frac{1}{4}$ ○ $\frac{1}{2}$

4. 6 hours 45 minutes
   +2 hours 25 minutes

   ○ 9 hr. 10 min.   ○ 4 hr. 20 min.
   ○ 9 hr. 70 min.   ○ 9 hr. 60 min.

Name _____ 5G-3

1. 75,881
   +63,097

   ○ 138,078   ○ 138,978

   ○ 161,118   ○ 12,784

2. $\frac{5}{6} - \frac{1}{6} =$

   ○ $\frac{2}{3}$ ○ $\frac{1}{2}$

   ○ $\frac{1}{3}$ ○ 1

3. Three multiples of 2 are:

   ○ 1, 2   ○ 0, 1, 2

   ○ 2, 4, 6   ○ 0, 2, 3

4. 1461
    x   4

   ○ 1465   ○ 6024

   ○ 5644   ○ 5844

Name _____ 5G-4

1. Three multiples of 5 are:

   ○ 1, 5, 10   ○ 0, 10, 20

   ○ 5, 10, 15   ○ 0, 1, 5

2. 8.31 + 9.6 =

   ○ 17.37   ○ 17.91

   ○ 16.91   ○ 1.05

3. Which symbol goes in the circle below?

   $$\frac{3}{8} \bigcirc \frac{1}{4}$$

   ○ >
   ○ <
   ○ =

4. A farm produces 60 bales of hay from one acre. How many bales of hay will 5 acres produce?

   ○ 300   ○ 12

   ○ 100   ○ 30

Name _____   5G-5

1. $1\frac{2}{9}$
    $+1\frac{1}{9}$

    ○ $\frac{1}{9}$ ○ $2\frac{1}{3}$

    ○ $\frac{1}{3}$ ○ $2\frac{1}{6}$

2. 340
    x 33

    ○ 1,360   ○ 11,253

    ○ 11,220   ○ 11,250

3. 5)559

    ○ 111 remainder 4
    ○ 112
    ○ 1011 remainder 4
    ○ 11 remainder 4

4. Three multiples of 4 are:

    ○ 1, 4, 8   ○ 4, 8, 12

    ○ 0, 1, 4   ○ 1, 2, 4

Name _____ 5G-6

1. $\frac{1}{2} = \frac{\square}{4}$

    ○ 1   ○ 3
    ○ 2   ○ 4

2. 88
    93
    +416
    ─────
    ○ 607   ○ 621
    ○ 597   ○ 521

3. 8)500

    ○ 72 remainder 4
    ○ 62 remainder 4
    ○ 602 remainder 4
    ○ 625

4. Maria reads 20 pages of her book each night. How many pages does she read in 4 nights?

    ○ 80   ○ 24
    ○ 5   ○ 100

Name _____ 5G-7

1. 510
    × 20
    ──────

    ○ 10,730   ○ 10,200

    ○ 10,100   ○ 1,540

2. 5 × ___ = 7 × 5

    ○ 12   ○ 35

    ○ 7   ○ 17

3. 5 hours 10 minutes
    −1 hour 20 minutes
    ────────────────────

    ○ 3 hr. 50 min.   ○ 4 hr. 0 min.

    ○ 6 hr. 20 min.   ○ 6 hr. 90 min.

4. $\frac{2}{3} = \frac{\square}{6}$

    ○ 3   ○ 5

    ○ 2   ○ 4

Name _____   5G-8

1. Which is the numeral five thousand, four?

   ○ 5400   ○ 5004

   ○ 5040   ○ 5044

2. Three multiples of 6 are:

   ○ 1, 3, 6   ○ 0, 1, 6

   ○ 6, 12, 18   ○ 1, 6, 10

3. $\frac{2}{6} + \frac{1}{6} =$

   ○ $\frac{3}{12}$ ○ $\frac{1}{2}$

   ○ $\frac{1}{4}$ ○ $\frac{1}{3}$

4.   600
     −129

   ○ 529   ○ 481

   ○ 729   ○ 471

Name _____   5G-9

1. $\frac{1}{4} = \frac{\Box}{8}$

    ○ 2   ○ 8
    ○ 4   ○ 6

2. $843.11
    -561.08

    ○ $1404.19   ○ $282.03
    ○ $382.13   ○ $322.17

3. $3\frac{1}{3}$
    $+2\frac{2}{3}$

    ○ 6   ○ $5\frac{1}{3}$
    ○ $5\frac{2}{3}$ ○ $\frac{2}{3}$

4. 6050 =

    ○ 6 thousands, 5 hundreds, 0 tens, 5 ones
    ○ 6 thousands, 0 hundreds, 5 tens, 5 ones
    ○ 6 thousands, 5 hundreds, 5 tens, 0 ones
    ○ 6 thousands, 0 hundreds, 5 tens, 0 ones

Name _____   5G-10

1.   95,061
     +37,783

   ○ 132,044   ○ 57,278

   ○ 161,114   ○ 132,844

2. $3\frac{5}{8}$
   $+\ \ \frac{2}{8}$

   ○ $3\frac{3}{8}$ ○ 4

   ○ $3\frac{7}{8}$ ○ $3\frac{3}{4}$

3. Helen was born in 1971.
   Larry was born in 1975.
   Who is older?

   ○ Helen

   ○ Sue

4. $\frac{3}{5} = \frac{\square}{15}$

   ○ 15   ○ 9

   ○ 6   ○ 10

Name _____   5G-11

1. $\frac{1}{2}$
    $+ \frac{1}{4}$

    ○ $\frac{1}{4}$ ○ $\frac{3}{4}$

    ○ 1   ○ $1\frac{1}{4}$

2. 19.74 - 13.63 =

    ○ 6.11   ○ 33.37

    ○ 6.01   ○ 33.31

3. 400
    x 2

    ○ 800   ○ 822

    ○ 402   ○ 820

4. $6\overline{)300}$

    ○ 5 remainder 6   ○ 500

    ○ 60   ○ 50

Name _____ 5G-12

1. The factors (divisors) of 15 are:

   ○ 0, 1, 3, 5   ○ 1, 3, 5, 15

   ○ 15, 30, 45   ○ 1, 2, 3, 4

2. 41,008
   -21,637

   ○ 19,371   ○ 62,645

   ○ 20,631   ○ 29,371

3. $94.00
   +56.18

   ○ $150.00   ○ $37.82

   ○ $150.18   ○ $140.18

4. $\frac{1}{2}$
   $+ \frac{1}{3}$

   ○ $\frac{5}{6}$ ○ $\frac{4}{6}$

   ○ $\frac{2}{5}$ ○ $\frac{1}{2}$

Name _____ 5G-13

1. $\frac{1}{3}$
    $+ \frac{1}{4}$

    ○ $\frac{1}{6}$ ○ $\frac{2}{3}$

    ○ $\frac{2}{7}$ ○ $\frac{7}{12}$

2. Find the average of these numbers:
    (10, 3, 6, 5)

    ○ 8   ○ 12

    ○ 6   ○ 2

3. 6 thousands
    7 hundreds
    9 tens
    0 ones =

    ○ 6709   ○ 6790

    ○ 679   ○ 7690

4. $5\frac{3}{8} - 1\frac{2}{8} =$

    ○ $6\frac{5}{8}$ ○ $4\frac{5}{16}$

    ○ $4\frac{1}{8}$ ○ $3\frac{3}{4}$

Name _____   5G-14

1.   23.9 + 41.20 =

   ○ 65.10   ○ 64.10

   ○ 65.29   ○ 43.59

2. $\frac{3}{4}$
     $-\frac{1}{2}$

   ○ $1\frac{1}{4}$ ○ $\frac{1}{4}$

   ○ 1   ○ $\frac{2}{3}$

3.   701
     x 24

   ○ 4,206   ○ 17,064

   ○ 16,824   ○ 725

4.   Betty's class is having a picnic. 40 people are going. 5 people can go in a car. How many cars are needed?

   ○ 8   ○ 4

   ○ 10   ○ 200

Name _____   5G-15

1. Three multiples of 10 are:

○ 1, 5, 10   ○ 10, 20, 30

○ 0, 5, 10   ○ 5, 10, 15

2. $6.23
    x   3

○ $6.26   ○ $18.79

○ $18.69   ○ $18.66

3. 735
   -217

○ 952   ○ 418

○ 518   ○ 528

4. 4)100

○ 25   ○ 250

○ 205   ○ 25 remainder 2

1. $68.43
        - 29.09

   ○ $39.04   ○ $97.52

   ○ $41.06   ○ $39.34

2. $\frac{1}{2}$
      $-\frac{1}{3}$

   ○ $\frac{1}{6}$ ○ $\frac{1}{3}$

   ○ $\frac{2}{5}$ ○ $\frac{1}{2}$

3. Round 430 to the nearest hundred.

   ○ 400   ○ 300

   ○ 500   ○ 450

4. $2\frac{4}{5}$
      $-1\frac{1}{5}$

   ○ 4   ○ $1\frac{3}{5}$

   ○ $3\frac{3}{5}$ ○ $1\frac{2}{5}$

Name _____   5G-17

1. 9 x (___ x 4) = (9 x 7) x 4

　　○ 63　　　　　○ 36

　　○ 9　　　　　 ○ 7

2. 　463
　　x 37

　　○ 4,630　　　　○ 16,131

　　○ 17,131　　　 ○ 500

3. 　51,120
　　-46,231

　　○ 97,351　　　 ○ 4,889

　　○ 15,111　　　 ○ 15,999

4. Tammy ate $\frac{1}{4}$ of the apple. Gary ate $\frac{1}{2}$ of the apple. How much of the apple did they eat together?

　　○ $\frac{3}{4}$　　　　　○ 1

　　○ $\frac{1}{2}$　　　　　○ $\frac{5}{8}$

Name _____   5G-18

1. 6.3 + 5.8 =

    ○ 12.1   ○ .5
    ○ 11.1   ○ 1.5

2. The factors (divisors) or 7 are:

    ○ 0, 1, 7   ○ 1, 7
    ○ 7, 14, 21   ○ 1, 3, 5, 7

3. $\frac{5}{8} - \frac{3}{8} =$

    ○ $\frac{1}{4}$ ○ $\frac{2}{0}$
    ○ 1   ○ $\frac{1}{8}$

4. 637
    x 4

    ○ 641   ○ 2548
    ○ 2422   ○ 2428

Name _____   5G-19

1. $\frac{1}{3}$
   $+\frac{1}{5}$

   ○ $\frac{2}{15}$ ○ $\frac{1}{4}$

   ○ $\frac{8}{15}$ ○ $\frac{1}{8}$

2. $\frac{3}{5}$ $\frac{3}{8}$ ⓐ$\frac{3}{4}$

   Of the fractions shown, the one circled has:

   ○ the greatest value
   ○ the least value
   ○ the same value

3. $4\frac{2}{3}$
   $-4\frac{1}{3}$

   ○ 9   ○ $\frac{1}{0}$

   ○ $\frac{1}{3}$ ○ $1\frac{1}{3}$

4. $56.42
     17.88
   +13.00

   ○ $76.20   ○ $87.30

   ○ $151.11   ○ $86.00

Name _____   5G-20

1. $6\frac{4}{6}$
    $-2\frac{2}{6}$

    ○ 9   ○ $4\frac{1}{6}$

    ○ $4\frac{2}{3}$ ○ $4\frac{1}{3}$

2. The earth goes around the sun once in 12 months. How many months does it take for the earth to go around the sun 6 times?

    ○ 6 months   ○ 18 months

    ○ 24 months   ○ 72 months

3. ___ x 9 = 36

    ○ 27   ○ 324

    ○ 45   ○ 4

4. 605
    x 45

    ○ 27,225   ○ 5,445

    ○ 27,005   ○ 27,072

Name _____   5H-1

1. Find the factors (divisors) of 9.

   ○ 2, 3, 6   ○ 0, 3, 9

   ○ 9, 18, 27   ○ 1, 3, 9

2. 16.23 + 24.28 =

   ○ 40.41   ○ 8.05

   ○ 40.51   ○ 30.51

3. $5\overline{)505}$

   ○ 911   ○ 11

   ○ 101   ○ 10 remainder 5

4. $\frac{1}{2}$
   $+ \frac{2}{5}$

   ○ $\frac{3}{5}$ ○ $\frac{9}{10}$

   ○ $\frac{1}{5}$ ○ $\frac{3}{7}$

Name _____   5H-2

1. 78,766
    -23,686

    ○ 55,080   ○ 55,120

    ○ 102,452   ○ 55,180

2. $\frac{1}{2} = \frac{\square}{6}$

    ○ 4   ○ 3

    ○ 2   ○ 5

3. 10)$\overline{20}$

    ○ 2   ○ 20

    ○ 1   ○ 11

4. Which sign goes in the circle below?

    80 + 0 ◯ 4 x 15

    ○ >
    ○ <
    ○ =

Name _____ 5H-3

1. $\frac{2}{3}$
    $-\frac{1}{2}$

    ○ $\frac{1}{2}$ ○ $\frac{5}{6}$

    ○ $\frac{1}{3}$ ○ $\frac{1}{6}$

2.   407
    $\times$ 80

    ○ 32,560   ○ 32,567

    ○ 3,256   ○ 32,650

3. A class left on a field trip at 1:45 p.m. and returned at 3:30 p.m. How long were they gone?

    ○ 2 hrs. 15 min.   ○ 1 hr. 15 min.

    ○ 1 hr. 45 min.   ○ 2 hrs. 45 min.

4. 2)$4.26

    ○ $2.13   ○ $2.01

    ○ $2.31   ○ $2.03

Name _____   5H-4

1. $3 \frac{1}{3}$
    $+2 \frac{2}{3}$
    _____

    ○ $5 \frac{3}{5}$ ○ 6

    ○ $5 \frac{4}{5}$ ○ $5 \frac{2}{3}$

2. Three multiples of 9 are:

    ○ 9, 18, 27   ○ 3, 6, 9

    ○ 0, 1, 3   ○ 1, 3, 9

3. $236.45
    - 118.19
    _____

    ○ $122.34   ○ $18.26

    ○ $118.26   ○ $128.36

4. 20)60̄

    ○ 2   ○ 30

    ○ 33   ○ 3

Name _____ 5H-5

1. 643
    x 71

    ○ 5,144   ○ 51,863

    ○ 45,653   ○ 51,763

2. 12)̄24

    ○ 4   ○ 2

    ○ 21   ○ 20

3. $\frac{4}{8} + \frac{2}{8} =$

    ○ $\frac{3}{8}$ ○ $\frac{3}{4}$

    ○ $\frac{1}{4}$ ○ $\frac{1}{8}$

4. 5000
    -4101

    ○ 899   ○ 809

    ○ 1101   ○ 1899

Name _____   5H-6

1. 4)$8.84

    ○ $2.02   ○ $2.01

    ○ $2.21   ○ $2.31

2. Grandfather caught 4 fish that weighed a total of 12 pounds. What was the average weight of each fish?

    ○ 3 lbs.   ○ 4 lbs.

    ○ 6 lbs.   ○ 2 lbs.

3.   561
       84
    +327

    ○ 962   ○ 972

    ○ 982   ○ 1072

4. $7 \frac{3}{8}$
    $-1 \frac{1}{8}$

    ○ $8 \frac{1}{2}$ ○ $6 \frac{3}{8}$

    ○ $6 \frac{1}{2}$ ○ $6 \frac{1}{4}$

Name _____   5H-7

1.   3 x 2 = 2 x ☐

   ○ 6   ○ 5
   ○ 7   ○ 3

2. $\frac{5}{6} - \frac{1}{6} =$

   ○ $\frac{4}{0}$ ○ $\frac{2}{3}$
   ○ $\frac{1}{2}$ ○ $\frac{1}{3}$

3.   3)$\overline{\$9.63}$

   ○ $3.21   ○ $3.12
   ○ $3.02   ○ $2.53

4.   6 hr. 10 min.
    -2 hr. 30 min.
    ─────────────

   ○ 4 hr. 20 min.   ○ 3 hr. 80 min.
   ○ 3 hr. 40 min.   ○ 4 hr. 40 min.

Name _____ 5H-8

1. $4.25
    x   6

   ○ $26.50   ○ $24.50

   ○ $25.60   ○ $25.50

2. $1 \times \frac{1}{3} =$

   ○ $1\frac{1}{3}$ ○ 1

   ○ $\frac{1}{3}$ ○ $\frac{2}{3}$

3. 20)‾200‾

   ○ 10   ○ 102

   ○ 100   ○ 12

4. 2051
    -1064

   ○ 1013   ○ 3115

   ○ 987   ○ 997

Name _____   5H-9

1. 5 thousands
    3 hundreds
    0 tens
    7 ones =

    ○ 5307   ○ 530

    ○ 7035   ○ 537

2. $\frac{2}{4} = \frac{\Box}{8}$

    ○ 2   ○ 4

    ○ 3   ○ 6

3. 420
    × 20
    ─────

    ○ 84   ○ 840

    ○ 8,400   ○ 84,000

4. 9 × (___ × 4) = (9 × 7) × 4

    ○ 7   ○ 16

    ○ 13   ○ 4

Name _____   5H-10

1. $\frac{3}{4} - \frac{1}{4} =$

   ○ $\frac{1}{4}$ ○ $\frac{3}{4}$

   ○ $\frac{1}{2}$ ○ 1

2. $12\overline{)240}$

   ○ 20   ○ 30

   ○ 202   ○ 20 remainder 2

3. $1 \times \frac{1}{2} =$

   ○ $1\frac{1}{2}$ ○ 1

   ○ 2   ○ $\frac{1}{2}$

4. $346.00
    − 125.68

   ○ $220.32   ○ $221.42

   ○ $221.68   ○ $121.68

Name _____   5H-11

1. 9.2 - 4.6 =

    ○ 5.4   ○ 13.8

    ○ 4.6   ○ 5.6

2. The first four factors (divisors) of 12 are:

    ○ 0, 1, 2, 3   ○ 1, 2, 3, 4

    ○ 2, 4, 6, 8   ○ 12, 24, 36, 48

3. $\frac{3}{4}$
    $-\frac{2}{3}$

    ○ $\frac{1}{12}$ ○ $\frac{1}{3}$

    ○ 1   ○ $\frac{5}{12}$

4. 8)$\overline{501}$

    ○ 66 remainder 5   ○ 72 remainder 5

    ○ 76 remainder 5   ○ 62 remainder 5

5H-12

1. $\dfrac{1}{3} = \dfrac{\Box}{9}$

   ○ 4   ○ 5

   ○ 3   ○ 6

2. $1\dfrac{1}{4}$
   $+2\dfrac{3}{4}$

   ○ 4   ○ $4\dfrac{1}{4}$

   ○ 3   ○ $3\dfrac{1}{4}$

3. At the School Carnival 30 quarts of cider were sold. One quart made 4 glasses. How many glasses of cider were sold?

   ○ 34   ○ $7\dfrac{1}{2}$

   ○ 100   ○ 120

4. $12\overline{)360}$

   ○ 300   ○ 30

   ○ 291   ○ 36

Name _____ 5H-13

1. 2)$2.64

    ○ $132   ○ $1.32

    ○ $130   ○ $13.2

2. Find the average of the following numbers:
    (6, 6, 14, 2)

    ○ 7   ○ 6

    ○ 8   ○ 5

3. 421
    x 32

    ○ 13,472   ○ 13,482

    ○ 2,105   ○ 9,683

4. $7 \frac{3}{8}$
    $-1 \frac{1}{8}$

    ○ $8 \frac{1}{2}$ ○ $6 \frac{1}{2}$

    ○ $5 \frac{1}{4}$ ○ $6 \frac{1}{4}$

Name _____   5H-14

1. Three multiples of 8 are:

   ○ 8, 10, 12   ○ 1, 2, 4

   ○ 8, 16, 24   ○ 8, 12, 20

2. $6\overline{)700}$

   ○ 16 remainder 4   ○ 116 remainder 4

   ○ 106 remainder 4   ○ 160 remainder 4

3. $\begin{array}{r}\frac{8}{9}\\+\frac{1}{9}\\\hline\end{array}$

   ○ $\frac{10}{9}$ ○ $\frac{7}{9}$

   ○ $\frac{1}{2}$ ○ 1

4. $\frac{3}{4} \times 1 =$

   ○ $\frac{3}{4}$ ○ $1\frac{3}{4}$

   ○ 4   ○ $1\frac{1}{3}$

Name _____ 5H-15

1. $\frac{2}{4}$
    $-\frac{1}{3}$

    ○ $\frac{5}{6}$ ○ $\frac{2}{7}$

    ○ $\frac{1}{12}$ ○ $\frac{1}{6}$

2. 23.9 + 41.21 =

    ○ 64.11   ○ 65.11

    ○ 43.60   ○ 18.12

3. $10\overline{)520}$

    ○ 52   ○ 50 remainder 2

    ○ 502   ○ 5 remainder 2

4. Russ cut 5 oranges into fourths. How many pieces did he make from the 5 oranges?

    ○ 9   ○ 25

    ○ 10   ○ 20

Name _____ 5H-16

1.   7010
   −1234

   ○ 5776   ○ 6224

   ○ 5886   ○ 5786

2. $\frac{5}{6} \times 1 =$

   ○ $\frac{5}{6}$ ○ $1\frac{5}{6}$

   ○ 6   ○ 1

3. $6\frac{7}{8}$
   $-\frac{3}{8}$

   ○ $\frac{1}{2}$ ○ $7\frac{1}{4}$

   ○ $6\frac{1}{2}$ ○ $6\frac{5}{8}$

4. Which numeral is sixty thousand, four hundred, twenty-eight?

   ○ 64,428   ○ 6,428

   ○ 60,428   ○ 600,428

Name _____   5H-17

The graph below shows the average monthly temperature in Indianapolis, Indiana, Use the graph to answer the questions that follow.

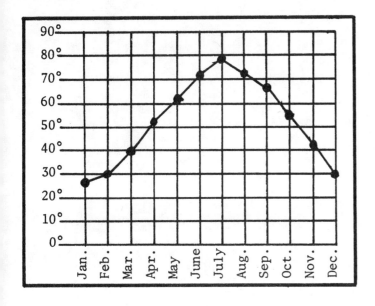

1. In what month is the average temperature the highest?
   ○ May   ○ June
   ○ July   ○ August

2. In what month is the average temperature the lowest?
   ○ December   ○ January
   ○ February   ○ March

3. What is the average temperature in March?
   ○ 30°   ○ 40°
   ○ 35°   ○ 50°

4. In what months is the average temperature below 40°?
   ○ Jan., Feb., Mar.   ○ Feb., Nov., Dec.
   ○ Mar., Nov., Dec.   ○ Jan., Feb., Dec.

Name _____   5H-18

1. 525
   x 25

   ○ 13,022   ○ 3,675
   ○ 550   ○ 13,125

2. Which list has only multiples of 7?

   ○ 1, 7   ○ 7, 14, 21
   ○ 7, 10, 15   ○ 7, 17, 27

3. $\frac{1}{3}$
   $+ \frac{2}{4}$

   ○ $\frac{3}{7}$ ○ $\frac{11}{12}$
   ○ $\frac{3}{4}$ ○ $\frac{5}{6}$

4. $26.83
    47.37
   +18.50

   ○ $92.60   ○ $92.50
   ○ $92.70   ○ $91.70

Name _____   5H-19

1.   30)960   2.   65.41 - 25.8 =

   ○ 302   ○ 3 remainder 6   ○ 39.61   ○ 91.21

   ○ 32   ○ 30 remainder 6   ○ 62.83   ○ 40.61

3.   A TV set has a 21-inch screen.   4. $\frac{1}{5} = \frac{\square}{10}$
     How much more than a foot is
     21 inches?
                                            ○ 2   ○ 4
   ○ 11 inches   ○ 10 inches
                                            ○ 6   ○ 3
   ○ 12 inches   ○ 9 inches

Name _____   5H-20

1.   2)$24.68

   ○ $102.34   ○ $12.34

   ○ $10.23   ○ $123.40

2. $\dfrac{\Box}{2} = 1$

   ○ 2   ○ 4

   ○ 3   ○ $\dfrac{1}{2}$

3.   4068
     x   2

   ○ 8126   ○ 8811

   ○ 8136   ○ 4070

4. $9\dfrac{6}{10}$
     $-\ \dfrac{1}{10}$

   ○ $9\dfrac{1}{2}$ ○ $8\dfrac{1}{2}$

   ○ $9\dfrac{7}{10}$ ○ 7